144 WAYS TO REDUCE YOUR

ORGANIZATION'S

ECOLOGICAL

FOOTPRINT

A guide for small- and medium-sized organizations interested in adopting socially and environmentally responsible organizational practices

plus 25 tips for employees who want to reduce their environmental footprint in the workplace.

Printed in the United States of America. For information, or if you want to purchase in bulk at discount rates, please visit www.humusamoris.com

Library and Archives Canada Cataloguing in Publication

Tourville, Michel, 1973-

144 ways to reduce your organization's ecological footprint : a guide for small- and medium-sized organizations interested in adopting socially and environmentally responsible practices / Michel Tourville.

Includes bibliographical references and index.

ISBN 978-0-9865456-0-3

1. Small business--Environmental aspects. 2. Business enterprises--Environmental aspects. 3. Environmental responsibility. 4. Social responsibility of business. I. Title. II. Title: One hundred forty-four ways to reduce your organization's ecological footprint. III. Title: Ecological footprint.

Preface

Humus Amoris was founded in 2008 after its founders came to the realization that they needed to contribute to ecological footprint reduction on a larger scale. This decision to try and have a meaningful impact on the greatest challenge the earth has yet faced meant that the cozy and comfortable life the author and his family were living needed to be challenged and adjustments made.

Through talking with acquaintances, friends, co-workers and relatives, we've come to realize that many small- and medium-sized organizations that are starting to become more aware of their ecological footprint don't know where to start when it comes to making ecologically-responsible business choices. It turned out that most organizations would be aware of one or two things they could do—or do differently—to make a difference, but that was pretty much the extent of it.

After doing a little research, we learned that it was hard to find books and web sites where smaller organizations could go to find much information in one place. Lots of web sites or references would offer tips on certain topics, but not many, and then would often provide too little or too much explanation along with technical elements that only eco-savvy people would find interesting. Also, very few site offered concrete examples of how their ideas could be applied to daily life. Most of the available studies being reported have focused on large companies or conglomerates that have very little in common with a five-volunteer local social organization.

After reflecting on our research, we came to the conclusion that the vast majority of people just want to contribute. Some of these people are against the environmental movement, while some are militant environmentalists, but as is true of most social causes, taking earth-friendly action is just good old common sense. People don't necessarily want the whole encyclopaedia thrown at them—they want to know where they can influence change, how they can act, and when they need to do so. They also want to know what the overall benefits of acting are so they can understand the role they play in the process and develop their own opinions. Our studies also show that people don't want to totally change how they live today. They want to *contribute*, but they don't want to commit to drastic measures, at least not at first.

So, while doing our research, we started compiling information we found along the way that we thought could be useful to people and organizations in reducing their ecological footprint. Our intent was to gather this information together in one easy to reference place. Then people could easily scroll through the different topics offered and pick tips they could easily put in to use, but at the same time be exposed to other ways of doing things that they wouldn't have though of on their own, with the hope that, eventually, they could change some of their "preferred" non-eco habits. Over the years, as people become more and more eco-aware, we notice that some of the tips here have become integrated into their lives, but the quest for ecological footprint reduction is a never-ending process.

Which leads us to the obvious question: Why 144 tips? Our answer? Why not! 144 is a great number!

But seriously, when we were selecting tips relevant to an organization, we ended up with 144. Squeezing them down to a nice round 100 would require mixing some content areas that we felt were best left separate, and watering down the tips so we could get to 150 just wouldn't be right.

Our intent is not to have a book with a ton of words, a ton of examples, or a book that repeats itself over and over just to increase page count. We've seen a few books being sold that spend 300 pages telling you things that could be explained in 20. See this as a book of ideas.

Our goal is to share are many tips as we can on a given subject and be as practical as we can so the tips can be understood and applied by anyone. Although they may help sell more books, you won't find celebrity endorsements here. You also won't find long case studies focusing on big companies or organisations that may make good reading but are really aren't pertinent or applicable to your organization.

On top of promoting eco-friendly events, activities, tips and products through its web site, Humus Amoris donates 3 % of its sales to green.

Contents

Introduction

The topic of the environment has come to the forefront of political and economic discussions over the last few years. More and more, a consensus of opinion is developing that people and organizations need to reduce their ecological footprint. With the population growing while natural resources are decreasing, it's just logical to see that we'll soon reach a point where there won't be enough resources on earth to feed and host mankind. Each of us must assess our footprint, how we live, and respect the environment.

So, what is your organization doing about it? Aside from maybe a few nice statements on your brochures or web site, what is your organization really doing that differs from the unsustainable, unacceptable way things are generally done?

Before explaining how you can use this book, let's start by discussing why your organization needs to move toward green practices and products.

The Global Context is Changing

By simply paying attention to what's happening around you, you can see that the earth is starting to struggle with its capacity to host the human species. As natural resources become rarer, Mother Nature is constantly reminding us of her fragile balance and that natural habitats are dying. You could say this problem isn't new, and you'd be right. What is new, though, is that people and their behaviors are starting to change. We see more and more companies not just content to comply with regulations, but instead being proactive and developing new methods that greatly exceed expectations.

Instead of having your organization focus on the bottom line, it now needs to focus on the triple bottom line as described in 1994 by John Elkington.

Element	Triple P
Humans	People
Nature	Planet
Finance	Profit

Each element is as important as the other. People are essential to organizations. They translate into healthy communities, good employees, good suppliers, and strong customer relations.

The second P (Planet) is obvious to you if you've bought this book. Sustainability must be a core element of a successful organization's mission.

Lastly, profit is essential for a business to survive, but it's also essential for non-profit organizations. Think of it this way: Even if the notion of profit doesn't exist within your non-profit organization, the notion of a balanced budget is critical to ensuring continuity. Elkington's strategy shows the connections between the three elements and long-term profitability and continuity.

When running a green organization, you need to understand what consumers and the public in general look for in a green organization. Some practices are obvious ones. Think of Exxon and the Exxon Valdes, for example. People use their pocketbooks to vote against organizations that have made the news for bad or eco-**un**friendly practices. The nature of an organization's products or services are also looked at by many. Of course, pricing has to be good, along with quality, but nowadays, consumers and potential clients also consider the type of packaging, its recyclability, and its impact on the environment once a product's useful life has ended.

Lastly, the origin of a product is becoming very important in the minds of earth-savvy consumers. There is a growing emphasis on supporting the local and regional economy which can't be ignored.

Be Green = Be Profitable

Operating as a green organization is a good way to improve your organization or small business' bottom line, since the first thing many organizations do when starting their green journey is to reduce their excessive consumption of various products or activities, which results in cost savings. Reducing waste and doing more with less are prime examples of this mindset and must become a *modus operandi,* or in other words, part of the organization's DNA.

Another financial benefit from instituting sustainable practices is improved revenues. When looking for ways to improve a product—by making it recyclable, for example—an organization is practicing innovation. This innovation may be a process, a product, or a service that's delivered differently than before, but can also be a means to distinguish your organization from the competition and gain market share. Another opportunity arises out of making long-lasting products. Sometimes when a second-hand use for a product is identified, it literally gives you another product which you can market to generate additional revenues. A good example of this is the wood residue found in mills, which we now see being repurposed into the compost and gardening mulch sold in gardening stores.

More and more, we see legal actions brought against organizations with regard to their treatment of workers or the environment. Activist movements are also stronger than ever before, and an organization's image can quickly become tarnished when caught in these situations. When building and maintaining a sustainable organization, all these aspects are considered during the decision-making process, thereby reducing the organization's overall legal exposure. A good example can be found in litigation on hazardous materials, which requires tremendous financial resources to administer, and, if not done, carries extraordinary potential financial penalties. So then why not invest in green products and natural alternatives, and eliminate that concern once and for all?

Although many organizations today declare themselves green, consumers are becoming more and more informed and are not fooled for long by organizations that talk a good game, but don't deliver. They gradually stop supporting those organizations who are green in name only. An organization must be true to its goals and transparent to its customers and partners. Over time, this will prove to be the right strategy.

In this day and age, green organizations are attractive not only to customers, but also to potential employees. High-performing and forward-thinking employees look for places to develop their careers and skills where they can work with others who share their goals and ideals. You, as a leader, know that the successful performance of your organization is not simply due to your ultra-sophisticated machines or that ultra-secret ingredient no one else uses, but is due to your employees, who make things happen, day in and day out. So, in order to capture new talent, encourage innovation in green subject matters, link sustainability with pride, and inject passion in your organization.

"But I Run a Municipal or Non-profit Organization."

Some elements influencing sustainable development are government-driven, and all levels of government, be they federal or municipal, have their say in how the game is played. Tax credits, loans, subsidies... these all influence project development. In this sense, governments can have a significant impact on how businesses operate.

As for non-profit organization, they have the potential to inspire business partners to undertake change. It's common for non-profits to influence businesses through association. Some non-profit organizations which have grown over the last decade and are now managing huge operating budgets have many opportunities to incorporate green practices in their everyday activities.

Bottom line: All non-profits have volunteers, processes, activities, and budgets, and can make an impact on the environment by how they manage these resources, however small they may be.

How You Can Use This Book

The intent of this book is to put as much information about as many easily-adoptable green practices as possible in one accessible place in order to help you or your organization take significant action to reduce your ecological footprint. Keep this book somewhere where you can easily access it, like your office or the coffee table. This book is not an encyclopedia where you'll find deep content and complex writing on an exhaustive number of subjects. It is intended to spark your interest and help you get started on your journey, even if you start with small steps. You may try some of the tips found here and find they're perfect for you, or and you may stop using them. No one is perfect, and any progress—even a small change—is still progress. Some tips may apply better in some situations than others, or be more applicable to some people and organizations than others. The worst thing that can happen is that things stay unchanged, which isn't a high price to pay for trying.

Using this book is simple. Just go though it by field of interest. Select a section that appeals to you (water consumption, for example) and read through it. Each section can be read in just a few minutes. Once you're done reading, choose one tip you can easily apply to your situation. Some tips won't require additional information to implement, but others may require that you do a little more research on the web or in books to gain a deeper knowledge and understanding of the subject.

Finally, just try to apply one tip at a time. When faced with all these opportunities at once, many people become frozen in place because they don't know where to start. A good place to start is to select the most easy to apply tip. Once this tip is mastered, go to the next easiest tip, and so on. Get in the habit of making changes one step at a time, as slow as it may seem.

Tips for Your Organization

The following sections contain tips to help reduce your organization's ecological footprint, organized by category. There are nine categories ranging from general tips, energy, habits, to consumption and so on. Each section is divided into subsections, with 11 subsections in total. These subsections will help you by offering more specific details about how and when to apply tips in a given area. For example, in the outdoor sub-section of the maintenance area are tips related to lawn-keeping, trees, shrubs, and so on. If there's a particular area you're interested in, we suggest that you use the Table of Contents to find your area of interest and start there.

Keep in mind, though, that the first thing you need to do before starting your journey to a greener organization is to measure your current ecological footprint to get a baseline and see where you're starting from. There are various web sites where you can go to get started, but we recommend the Greenhouse Gas Protocol web site at: www.ghgprotocol.org. We also suggest that you do this exercise in consultation with an experienced carbon consultant if possible. An expert will help explore the details or your particular situation. If your organization is located in a large metropolitan area, try a well-known accounting firm such as Deloitte & Touche, Ernst & Young, Pricewaterhouse Coopers, or KPMG—chances are they have, or have access to, sustainability experts close to you. By gauging your current situation, you'll be in a better position to see the improvements you make as your make them, which is not only motivational, but will help when you start communicating with the outside world about these changes, whether for marketing purposes or just information. You'll be able to show that you have real results to show, and are not just blowing hot air.

General Tips

1. Adopt a Green Mission or Policy

Your mission statement is where everything starts. The environment needs to be part of why your company exists. Obviously, you want to manufacture products, delight your customers, give outstanding service, but it's all worthless, over time, if you don't act in ways that respect the environment. Your mission is public—customers see it, suppliers see it, and most importantly, employees see it. So if you want to put environmental policies in place for your organization, or implement new ways of doing things in regard to the environment, it's much easier to take action if you can relate these actions to your organization's mission and make these actions part of your organization's values. This will also pressure you, in a good way, to adopting certain new practices if your mission has an environmental twist to it. We're not saying to have a mission that's green all over, but even a few simple words like, "In respect of the environment...," are a great start.

It's now been proven that organizations with sound environmental management policies perform better than their less environmental counterparts. The main reasons for this are that organizations that consider these things are ready to face global challenges and are well-positioned in this new economy.

2. Practice Green Management

The environment needs to be at the center of your working culture and needs to be part of your organization's mission. Corporations are now creating positions like Vice President or Chief Environmental Officer in order to address environmental issues. For smaller organizations that are not in a position to hire someone dedicated solely to the task, give this responsibility to someone with a title central to your operations who will impact as many employees and activities as possible — Purchasing Manager, for example, if procurement is a big part of your organization's activities and represents a good portion of your workforce. The idea is not to create a structure just for the sake of it, but to put focus on eco-actions and demonstrate to employees that the environment is key to your organization's strategy.

While the manager you recruit must have the basic skills required for the job, also consider other personal traits that will make this person a successful sustainable thinker. The first thing you may want to check is a potential employee's curiosity about environmental practices. Have they read green literature lately? Are they aware of current green business trends, for example? The person you entrust this position to also needs to have a strong sense of justice. In order to run an organization respectful to both the environment and the workforce, they will need to pay as much attention to the human elements involved in their decision making as any other elements. Having strong principles that align with your organization's mission and goals is also key. On this journey, managers will feel alone in their positions and convictions more times than not. Perseverance is essential to promote green changes.

Once you've evaluated personal traits, there are also a few business skills that can be very useful to look for when recruiting a green leader. Risk management skills are important to have. We're used to hearing about risk management in the sense of profits and losses, but remember, a green manager works to increase the triple bottom line. Look for someone willing to take a risk by confronting senior management on their green policies, or proposing and pushing for new policies that will be good for employees and their well-being. In this day and age, green is still not what we could consider mainstream, unfortunately, so you need good sellers to help you change that. These new managers will need to convince people to adopt a new way of doing things and bring people together on the subject matter. The green manager must also have a 'continuous improvement' mindset. In the green world, you can't make a change and consider it as a check in a box. Organizational sustainability is a never-ending process where each action and decision, as small as it may seem, is a step toward a better balance between the 3 P's: People, Planet, and Profit. Different elements (external or internal) can affect that balance and a good manager both reacts and acts to maintain that balance at all times.

Finally, green managers must be able to really understand the big picture in order to make the right decisions, and they must be able to think outside the box. In the green world, sometimes the solution is not known, because the problem you find yourself facing was never handled the green way. So, you start by reducing the impact of your problem by placing mitigation plans in place, but the true resolution will come from innovation.

3. Setting Green Goals

Setting goals is a way to put your thought processes into action. Goals are catalysts to change. To be effective a good goal needs to be SMART.

S: Sustainable • There's no point in doing something that will only work once. Remember, one of the reasons you're implementing a green action plan is for sustainability.

M: Measurable • You can't assess what you don't measure. Make sure the goals you formulate have distinct, quantifiable performance indicators that will clearly measure what you want to accomplish.

A: Achievable • Don't dream. Or *dream*, but be honest with yourself with respect to the resources you can assign to your goal, whether human or financial.

R: Realistic • Let's face it, you can't and won't do everything at once. It's important to have a vision and to stretch goals, but don't discourage your contributors. If you start from a place far from the place you envision, break out your objectives in phases that are more in reach.

T: Timely • In your goal formulation, include concrete deadlines, such as, "within 12 months," or "every month." This will help communicate you goals to others and clarify expectations about when goals need to be met.

4. Green Key Performance Indicators

Key performance indicators, commonly known as KPIs, are a series of metrics that your organization will use to measure its progress toward green objectives. You manage what you measure. Set your goals and start to measure accomplishment of that goal. Try to have at least one goal for the main categories, such as water consumption, air emission, solid waste, energy, raw materials and packaging. It's good to put policies in place or hold meetings on how to improve your organization's situation, but in the end, what matters are the facts: The *true effect* your organization has on the environment. KPIs will show whether or not you're progressing toward your goals, so you can take appropriate actions.

5. Donate to the Cause or Join an Association

Donating to a cause your organization believes is one of the more cost effective ways to associate organizational objectives with your dedication to the environment. It's now common to see organizations donate 1% of their sales to a favoured cause, so if you want to distinguish yourself from the masses, aim higher. Some sectors may not support high enough profit margins to support this type of donation, but the important thing is to start somewhere, as small as that effort may seem. Donation dollars will eventually add up over time, and if you've absorbed the initial amount set aside without too much injury to your bottom line, you may decide to increase your donation percentage over time. Groups like 'One Percent for the Planet' (www.onepercentfortheplanet.org/en/) promote this philosophy and encourage sustainable practices. You can also find organizations you can donate to at the website listed above.

Small organizations often become more powerful when they team with others who share their same interests and values. Today, many recognized associations have a green policy or are starting to look at greening practices. To find out, go to your current or targeted association's website and look around. Associations are usually proud to mention their involvement.

6. Celebrate Earth Day

Mark down this date: April 22nd. Be a part of or organize any event you possibly can. Bring your employees, friends, and family, and have your children participate, too. Make some noise! People need to see and celebrate Earth Day. This is also a good opportunity to reflect on your past environmental efforts throughout the year, think about the actions you have taken on this journey, appreciate the earth's beauty and pleasures, and congratulate your organization and your employees on their dedication to and involvement in the cause. Visit http://www.earthday.net to get ideas about various celebrations.

7. Social and Local Engagement

Nowadays, it's not just expected to donate money to causes, but also to get involved by donating time, resources, or knowledge. Being involved in your community enhances your organization's reputation, increases your customer's loyalty, and helps in recruitment and retention of the best employees.

Base your relationships in that regard on mutual exchange. Bring your expertise to these organizations, such as planning, if you excel in that area. This will also teach you a little about dealing with different stakeholders.

Time is also a resource that you can donate. There are a numerous local groups that can benefit from your support. Make a commitment within your organization to a culture of volunteerism by encouraging and supporting employee efforts to help out social projects during their work week.

Local groups may be engaged in various activities, such as planting trees, recycling, or energy, and you can find them by going to the National Wildlife Federation's Conservation Directory (www.nwf.org/conservationdirectory). Make participation part of your employees' annual reviews so that participation is measured and has meaning.

8. Recycle

Did you know that a normal person can easily reduce his or her waste by 75% when recycling the basics (paper, glass, aluminum)? Think about the impact of recycling on landfills and the natural resources saved when using recycled materials in new products or services. If your organization doesn't recycle, start your recycling program now.

To begin recycling, you'll first need a leader to develop and implement your recycling program. Try to look for someone who is passionate about the topic. A common place to look is among your facilities people, since they already see what gets put in the garbage and where your recycling opportunities may be.

Next, identify what size recycling containers you'll require. Volume of recyclable objects, level of activity in a given area, and the frequency of collections are all parameters that need to be evaluated.

People will also need to be trained to separate their things. Cover the basics to start with, but also develop a more formal and detailed guide that people can consult when needed. Pictograms on bins are very important as people will get to the recycling area and then ask themselves where to put their objects. If the pictograms on the bins are not clear, chances are items will end up in the wrong bins.

Like they say in real estate, the three most important things for a recycling program are location, location and location. If your bins are not well placed, people won't go out of their way to help your program. Processes need to be easy for the users. The lunch area will be the first thing that needs to be covered, but if you can place baskets at each workstation, this is also great. Don't go overboard, too many bins will result in a waste of space and higher collection costs. On the other hand, not enough bins will result in overloaded bins and lack of employee participation.

Once the basics of your plan are in place, you'll need to figure out who will do rounds to collect the bins, how often they'll need to be picked up, and how you'll need to synchronize this with the material leaving your facility for the recycling centers. This last point is important as, depending on where you live, some of your material can likely be picked up by the weekly city collection truck, but other materials may need to bundled or arrangements made for them to be transported to the closest recycling center.

9. Organize a Tree Planting Day Once a Year

Once a year, organize a workplace tree planting day. Make it a fun day for the family and kids by bringing food or organizing games. This event will also be a good time to get to know the people you work with better. After all, you spend as much time with them as you do with loved ones. Contact your local municipal leaders and see if they're willing to contribute by paying for the trees or giving you a place to plant them.

10. Set Green in the Work Culture

Empower your employees to not only participate, but also to make changes. Reward and encourage employees who have made a notable green contribution. If you're an environmental leader (and you're reading this book, aren't you?) set an example that others will follow. Set-up environmental goals and metrics that your organization will be measured on, such as water and energy consumption, waste disposal, recycled materials, etc. Communicate what you expect in regard to green behaviours and train all new employees on how they can participate in your waste reduction program(s).

When selecting people for green projects, one character trait to look for is passion. Without it you won't get the best out of people. Ideally, green team members need to be as concerned with your organization's potential on green matters as with their own personal wealth or gain. Try to tie the two together.

You also want green team members that represent all areas of your organization. One way to look at this is by looking at your organizational chart, then trying to select positions (not people) that would be helpful to have actively participating on your team. Considering this process in terms of employee position rather than by individual employees will provide better continuity if the original person leaves the team or company, since their replacement will automatically become a member of the team. If your organization is big enough, ask department managers to nominate someone to the team and if no one volunteers, require managers themselves to participate.

Depending on your organization, culture, and so on, it's not guaranteed that people will voluntarily get in line to join your green team. Try to make it easy for people to understand your project or initiative. Tell them they don't need to be experts, that decisions will be made as a team, and that they will not be put on the spot if they're not comfortable in the position. You also need to make your project enticing. It needs to be relevant to the organization's growth and health, so that in return it will benefit employees, for instance through enhanced job security.

Executive or high management sponsorship and participation is important, too. This brings credibility to your initiative and shows that it is supported. If you have trouble recruiting an executive to the task, or don't have any, why not approach a board member who demonstrates environmental interest?

In terms of rewards, it's been proven that bonuses awarded based on quarterly results can prevent people from seeing what creates long-term prosperity. Link salary and bonuses to achievement of long-term environmental, social, and governance goals.

11. Replace Traditional Gifts

One good way to reduce waste is to offer non-tangible gifts. Organizations commonly offer souvenirs like mugs, pens, and t-shirts at Christmas and other special occasions. Why not replace this with tickets to a show, gift cards for a meal or a massage, and use electronic cards instead of paper? These are things your customers and employees will really appreciate while creating no waste. They're also a great ways to reward or encourage employees. If you still feel you need to give something solid, why not encourage products that respect your values, such as organic wine or chocolates?

12. Favor Green Companies and Invest Wisely

Keep in mind that your purchasing power is a means to influence the organizations you work with. The Environmental Protection Agency (EPA) tracks how companies perform environmentally with their National Environmental Performance Track. This rating is a good way to gauge which companies are responsible so you can encourage them with your spending power rather than their less eco-friendly competitors.

Making ethical and socially responsible investments plays a key role in encouraging organizations that have good environmental practices while re-routing money away from problem sectors. If your organization has an employee savings plan, you can structure it to offer these choices.

When selecting new suppliers, don't just pick the cheapest one. Often, low-cost suppliers can afford to bid more cheaply because they make their goods overseas and often operate in countries where regulation is less rigorous. While this sort of bid may be lower, the cost to the environment is high. By adopting responsible hiring policies with respect to suppliers, you also encourage the organizations that are the backbone of your local community. This creates more jobs locally, and that results in more money being reinvested locally

13. Uniforms

When selecting a supplier for uniforms, don't neglect to choose one that ensures your company's moral comfort through the materials used and the way they run their organization. Synthetic fibers are non-biodegradable. Even cotton is problematic, since cotton is the world's most chemically-intensive crop, requiring up to 18 applications of herbicides and insecticides during the growing process. Wool also requires a great deal of water to process—over 20,000 gallons per pound. The best fabric options are chemical-free organic cotton and hemp.

14. Product Design and Output Reuse

If you sell or manufacture a product, try to adhere to good environmental practices throughout the process. Buying locally when possible, using organic or recycled products, and communicating electronically instead of via paper are all ways to reduce your product's ecological footprint. Remember, in the green world, when you're bringing a product to market, you should always think about the whole life of the product from cradle to grave. Don't encourage planned obsolescence to ensure continued demand at the cost of valuable resources.

Create products that can be repaired, upgraded, reused and recycled.

It all starts with your choice of materials. Try to use the least virgin raw materials you can, and select non-toxic and bio-based materials when possible. Then, design your product as simply as possible by avoiding unnecessary bells and whistles which make your product harder to manufacture (processing, machine time, etc.) but don't bring value to the customer. Always design for long product life. Stay away from short-term trends. Have your product designed so it's long-lasting and can be reused for other needs. Lastly, consider how your product will perform environmentally during its life. Will it burn a great deal of fuel or consume a great deal of water, for example?

In any case, you don't need to reinvent the wheel. Resources are already available. For example, Engineers for a Sustainable World promotes sustainable thinking for new engineers. They hold conferences and are present on many university campuses. To find out more, go to: www.eswusa.org.

Another organization to look into is the Biomimicry Guild. This organization conducts workshops, provides consulting, and coordinates research to help product developers think about design by examining ecosystems. Check it out at: http://www.biomimicryguild.com/.

Many organizations manufacture a great deal of unused output. This output can be the heat generated by machinery, or processing residue such as metal chips or cooking oil. In your operations, think about how you can use that output. Can the heat generated by your machines be used to heat water? Can you sell old oil so it can be turned into biodiesel? You might also consider developing partnerships in which you make arrangements with other organizations that can use your leftover products. Can you recycle those metal chips? In any case, think of what gets wasted and see if you can't make the best use of it.

15. Green Finances

Greening your finances will provide many opportunities surrounding sustainable development. Locating your finance activities on the green side doesn't happen overnight, but as with other tips in this book, start slowly, keep moving in the right direction at a steady pace, and eventually you'll surprise yourself with how much you've progressed.

What is green financing? Green financing is the process of handling your traditional finance activities with the triple bottom line in mind with regard to operations and policies. Having green financial practices will give you access to a new circle of people able to help your organization in its activities. This will put you in contact with other business people that may think as you do with regard to the environment. Here are a few things that the finance department can do to green your financial practices:

- Provide bank and financial statements online;
- Implement greener strategies;
- Provide financial support for green projects;
- Institute a matching funds policy with respect to employee charitable contributions;

- Choose to do business with green financial institutions;

- Apply for green mortgages; and

- Choose local investments.

Another thing to consider is using green credit cards. Green credit cards are usually made out of corn and offer ways to incorporate sustainability practices. They can offer you participation in green power projects, buy carbon offsets, or donate to environmental causes. The Brighter Planet and the GreenCard, both offered by Visa, are two examples of green credit cards.

Certifications and Labeling

16. ISO 14000

The International Organization for Standardization has developed methods for environmental assessments which cover management, labelling and certification. Knowing what these certifications mean will help you in your daily decision making regarding purchases, and help you decide what type of certification you want your products to carry. There are three levels of certification:

Level 1 (ISO14024): Denotes selective, multi-criteria-based, third party certified endorsement of a product. This is the most valuable certification of the three for both producers and consumers.

Level 2 (ISO14021): Level 2 is a self-declared claim, and as such is obviously the least meaningful of the three certification levels due to its subjectivity.

Level 3 (ISO14025): Level 3 certification provides quantified but non-selective information on the product based on independent verifications established against industry benchmarks.

For more information on ISO certification, please see: www.iso.org

17. Green Seal

Green Seal labels (greenseal.org) certify products that are environmentally better, from cradle to grave, than other products in the same category. Independent auditors award this label. Criteria are based on environmental quality and in conformity with ISO 14024 standards for Type 1 labels. This label is specifically used for office products.

18. Forest Stewardship Council

Some lumber imported into the U.S. is harvested illegally. The Forest Stewardship Council (FSC) identifies forest products containing FSC materials which assures these materials have not been harvested illegally. Standards used for assessment are agreed on by the FSC, which is a coalition of buyers, traders and non-governmental organizations drawn from the timber industry. This organization includes recognized groups such as the WWF and the Nature Conservancy. For more information, consult: http://www.fsc.org.

19. Donate and Recycle DVDs

DVDs are often used in the workplace to store software and data. In many cases, information becomes obsolete but disks remain and accumulate dust. If you do a clean-up of your DVD collection, bring them to a DVD recycling center. Not all recycling centers can handle DVDs, so make sure you check with your recycling center on their policies.

20. Envelopes

The obvious thing to do when you get bills or letters is to recycle used envelopes once opened, but you can also give these used envelopes a second life by using them as draft paper for to-do lists, grocery lists, or by keeping a stack close to the phone in case you need to note down a phone number. You can also encourage employees to use them for scratch paper at home. When folded, they're the perfect size to put in your back pocket, plus they're more rigid than common sheet paper so you can easily check things off without needing support.

21. Green Suppliers

When gathering bids for supplies needed by your organization, be sure to incorporate sustainability requirements. A simple example is to ask your cleaning services bidders to use green cleaning products.

Another thing to look at is proximity. Buying or using local products requires less transportation and therefore less gas, but also helps to build your regional economy.

22. Green Committees

Have your employees participate in a green committee. Help create a green organisational culture with a management plan that gets everyone involved. Let employees come up with ways to act in greener ways and give them the power to implement these changes—assuming they don't drive you out of business in the process! Ensuring that everyone has easy access to recycling collection bins is a good start, and adopting energy reduction policies is another great practice. Like any new project, begin by asking yourself what could be done to reduce your office's ecological footprint. Once you've chosen a way, plan for it. You'll need to communicate with your employees, but also consider whether on not they'll need training. Do you need to purchase things? When is the best time to implement? These questions need to be answered in order to ensure good planning. Good planning, in turn, increases your chances of success. Next, you'll need to execute you plan and, finally, measure your results. Keep track of what should be reduced, whether it's energy or waste. If you find that results are not as you had planned, go back and reassess you plan.

23. Correspondences

First, make sure you recycle unwanted correspondence, but more importantly, make sure you advise senders when you don't want to be on a mailing list. This also holds true for electronic documents, as they require time and disk space to manage. Think of how much time is wasted in your organisation in receiving, handling, reading, and getting rid of unwanted mail. It's surely not a productivity enabler.

24. Computers, Monitors, and Printers.

Computer and computer monitors have sleep or hibernate options which trigger when they go untouched or unused for a pre-defined length of time. These devices only use 5% of their full power in sleep mode. Develop the good habit of turning off all unused electronic devices when you leave work for the night and weekend to avoid using unnecessary energy. You should also make a habit of turning off computers and monitors when you know you're headed into a long meeting or out to lunch. Leaving these devices on generates heat which requires your office air conditioning systems to work harder. In some offices with many devices, their heat output can account for up to 30% of the air conditioning requirements of a given space. Nowadays, when new buildings are built, professionals need to factor equipment usage into their energy need calculations along with the activity that will take place in a specific room, otherwise air conditioning systems may not be powerful enough to compensate or heaters may be too powerful.

When buying computers, make sure you consult the Green Electronic Council's Electronic Product Environmental assessment tool, which rates computer makers on material use and their end of life take back policies. Also, try to favour companies which supply devices that can be upgraded and repaired to increase their life span.

About 75% of all computers sold in the U.S. end up in landfills or stockpiled for future disposal. This contributes to enormous lead, cadmium and mercury concentrations in those places. When your computers are no longer needed, donate them to schools, local organizations, or organizations that can reuse their parts. Check out the Basel Action Network (BAN) web site (www.ban.org), which provides a list of qualified North American recyclers. Many of these recyclers will offer mail-in services for electronics.

According to the United Nations (UN) environmental program, approximately 50 million metric tonnes of electronic waste are discarded every year throughout the world. Almost 90% of this waste gets exported to China and Africa, where it becomes subject to the manual and very dangerous practice of collecting materials that still have value (aluminum, copper, silver, etc.)

25. Default Double-sided Printing

You'd be surprised to see how much paper printed on only one side ends up in the recyclable bin, or worse, in the garbage. When ready to dispose of such paper, put it in a pile and use it again on the other side for draft print jobs or simply use it as you would a notepad. Set all your printers and photocopiers to print on both sides by default.

26. Ink

People often get fooled by the price of very cheap new printers. Ink isn't inexpensive, though, and you can easily get carried away with all those great colors. Set your printer to print in draft mode by default, which saves about 50% more ink. Most of what you print is for yourself rather than publication. Avoid going overboard with colors, too.

27. Internal Paperwork

Use electronic documents instead of paper for time sheets, expense reports, claims, or any internal-use documents. This will save on paper, plus it helps you understand these processes better because, once stored electronically, you can extract data that might be interesting to know or help better manage your organization. If you must use paper, buy recycled paper. Production of recycled paper uses up to 90% less water and half the energy required to make paper from lumber.

28. Paper and Ink Reduction

When personal computers were invented, we were told we'd eventually be working in paperless environments. The opposite happened. It's become so simple to correct a document and re-print it, as opposed to re-typing it on a typewriter like we used to do, that people can print 10 copies of what used to take two to three to get right the old-fashioned way. Make paper and ink use reduction one of your key environmental objectives for the year and give yourself an aggressive target. People should think twice before printing. This would not only save trees, but also storage, handling and fuel. Print on both sides, have all important documents spell-checked before printing, and when doing a presentation that absolutely requires you to print handouts, print at least three slides per page and use both sides of the paper.

You can also save paper by reducing your documents' default margins—try ½ inch all around. This needs some getting used to, as there is more text per page, but can help you save up to 20% in paper.

There is a software program called *GreenPrint* which is designed to eliminate unnecessary pages and converts files to PDF for sharing electronically. *GreenPrint* also calculates your savings so you can track and see your reduction in number of pages. Another good practice is to ban high-gloss products and color papers which are not recyclable.

Finally, for the paper that you must buy, look for unbleached post-consumer waste paper. For virgin paper, look for Forest Stewardship Council (FSC) certification to ensure the wood used was harvested responsibly.

29. Recycle Ink Cartridges

Spent ink cartridges take up a great deal of space in landfills. Aside from the warranty from the manufacturer, there is no good reason why cartridges can't be recycled. More and more re-fillers offer written guarantees against down time and equipment damage, so ask for these. Finally, recycled cartridges can cost 90% less than new ones—an incentive most managers can appreciate.

30. Collect Cell Phones

Provide a bin to collect old cell phones and send them to organizations like ReCellular, which collects and reprocesses more than three million retired phones annually. Choose a location in the organization where most people pass by to make it easy for them to deposit the phone. If you deal with a service provider for your mobile phone services, ensure they apply this policy also.

31. Reusable Mugs vs. Cardboard

Thank God Styrofoam cups are disappearing from our landscape. The next step is to use ceramic or metal mugs. In the office, when people need to pay for coffee, have disposable cups removed and provide a financial incentive for people to bring their cups back. Life cycle analysis shows that over its average life of 3000 uses, a mug generates 30 times less solid waste and 60 times less air pollution than a foam or cardboard cup.

32. Coffee

Change out the coffee you have at work in favour of organic and Fair Trade-labelled brands. Fair Trade certification promotes more sustainable agriculture. Fair Trade distributors pay a higher than market price for products to ensure minimum labour, environmental, and social conditions are met. For more information on fair trade certification, go to: www.sustainableharvest.com.

33. Coffee Stirrers and Milk

Replace coffee plastic coffee stirrers with dried pasta such as fettuccini. The cost is similar, pasta doesn't alter the coffee's taste, and pasta is 100% biodegradable. Instead of keeping those small cups of cream or milk, place a milk dispenser near the coffee machine, or if your break area has a refrigerator, just have a carton of milk available in the refrigerator.

34. Coffee Machine Filters

Make sure your office's coffee machine has a metal or plastic filter. This will save you from using paper filters every time the coffee machine is used. While you'll require more water to clean the filters, this is still less water than would go into producing those paper filters, plus it also creates less waste in landfills.

35. Office Kitchens

Allow your employees to bring their food from home. This habit is cheaper for the employees and produces less solid waste. Your office's kitchen should have recycling bins for glass, plastic, aluminum recycling, and a compost bin, if possible, even if this requires some spending. Don't offer plastic or paper cups and utensils—if they're available, people will use them. Provide real dishes, mugs and utensils. In the long run, you will save money, also.

36. Bottled Water

Avoid buying water bottles. Although recyclable, the quantity of bottles produced and not recycled is becoming incredibly high in landfills. At work, have your people make habit of filling their personal bottle. Standards defined by the EPA are more stringent than the standards set for bolted water by the Food and Drug Administration. If you want to be more drastic, remove bottled water from your organization's vending machines. For safety measure, educate workers on the importance of cleaning their personal bottle after each use in order to eliminate bacteria in the bottle. If you're unsure about your building's water quality, or if people don't like its taste, try faucet filters.

37. Don't Drink Traditional Soda

Replace traditional soda brands in your vending machine with the new brands of organic soda available. Consider club soda, which is made from filtered water that's been carbonated. This way you also avoid drinking artificial flavours and preservatives. Some brands to try include Santa Cruz Natural or Blue Sky Beverage Company.

38. Automatic Checks Deposits

While direct deposit is widely implemented, there are still people and small enterprises that give out paper pay checks every week. Converting to electronic pay checks not only saves paper but also saves employees time (don't need to cash it), fuel, and money.

39. Green Office Supplies

Almost anything used in an office can be somewhat green. From promotional material to the paper used in the printers, there's no limit! Below you will find supplies that can be easily switched out in favour of greener choices. Choose a source for office supplies located as closely as possible to your organization. Look for products manufactured in your country or continent that will require less transportation.

40. Adhesive Notes

Before having all these notes spread over your desk and computer, ask yourself if you really need this type of supply. Do you really need the combination of adhesive and paper, or can paper alone do the job? If you do buy adhesive notes, buy ones made out of recycled paper and that contain post-consumer content. Sometimes post-consumer content isn't always indicated on the label, but ask for it. The more consumers ask, the more corporations will adjust their materials strategies.

41. Binders

First, always use your binders more than once. If you no longer need the contents of a binder but feel you need to keep it, store it in a cabinet or in properly labelled storage boxes. Every year, check to see which ones are still good and which ones need a little repair, then try to find binders made with recycled materials.

42. Pens

Instead of buying cheap pens and throwing them away when there's no ink left, use a long-life refillable pen made out of recycled plastic, paper, or timber. Why not rely more on electronic means to write down notes? It's now a common practice to bring a computer to a meeting to take down notes. You'll save paper, ink, and time since you don't need to re-type the contents to send minutes or actions via e-mail. If you're more the pencil type, buy pencils made with wood off-cuts or recycled paper.

43. Paper Clips

Reuse, reuse, reuse. If you look inside office drawers in North America, you're sure to find paper clips. Everyone has them. Set aside a reclamation area where people can leave their unneeded clips and others can take them as needed.

44. Scissors

If you need new scissors, it's either because your old pair is no longer working efficiently, or because you don't have any. If you don't hair a pair, only buy scissors made with recycled stainless steel blades and plastic handles containing post-consumer plastic. Some pairs contain as much as 70% recycled materials.

If your old scissors don't work properly anymore, see if you can have them repaired. Maybe the blades just aren't sharp anymore. If so, have them sharpened. If there's a problem with the screw in the middle, go to your local knife store—there's a good chance your scissors can be able to be repaired.

45. Tape Dispensers

Try to buy tape dispensers made of at least 50% post-consumer plastic, and make sure if yours is broken or damaged to put it in the recycling bin. Not all recyclers take them, but the more your put them in the bin, the more recyclers will see there is volume and will start taking them.

46. Waste Baskets.

Choose waste baskets made from recycled steel, which takes less energy to produce than products made from iron ore, for example. Baskets made out of recycled plastics are also a good option.

47. Avoid Markers with Solvent Based Ink

If possible, replace solvent-based markers with chalk, wax pens, or simple color pencils. There are plenty of options when you spend some time thinking about it.

48. Use Fewer Napkins

In cafeterias and public eating places where napkins are offered for free, people have a tendency to take more napkins than necessary because they're afraid of needing to get back up to get a few more. Try to educate them about all that waste. Maybe you can make the napkins a little less available by placing them under the counter, for example. Or maybe people will need to ask for them. This will make people be more reasonable when taking napkins. If your organization is small, why not provide hand towels that you have washed by a cleaning service every day?

49. Consume Less

You've heard the proverb that more is better. This goes against the ecological footprint reduction concept. Today, we as individuals and organizations consume far more food, equipment, and gadgets than we actually need. Before making a purchase for the workplace, ask yourself if this product is really needed now, at this moment, or can wait. If the answer to is that it can wait and you're able to postpone that purchase, chances are you won't buy it in the end. Your organization's immediate apparent need will have passed or moved on to something else, and you'll avoid making unnecessary purchases which is better for your organization's cash flow and better for the environment.

50. Consume Right

It's often said in our capitalist world that the only good thing a consumer can do is buy, buy, and buy again. Today, with environmental concerns in mind, consumers and corporate citizens need to buy products or services that will be produced or delivered in a manner respectful of the environment and the people associated with it.

51. "Just in Case" purchases

Ever bought something "just in case", only to realize that you now have three of the same product, or that the co-worker seated next to you has one that he can share? If you're unsure, postpone your purchase. If it turns out you really need it, you're probably sure, not uncertain.

52. Buy Organic

Buying organic food for your organization not only ensures that you, your employees, or customers are eating food that's produced without pesticides and so on, but it's also a statement about sustaining durable development. Respect the soil, the air, and the people involved with these products. The U.S. Department of Agriculture's national organic program includes a set of standards verified by an independent party. Organic food is now available in most supermarkets and is used in restaurants and cafeterias.

53. Batteries

The number of batteries your employees go through in the course of the year can be astonishing. Start a recycling program at work. Your recycling program can be very simple. Here's a trick: Use an old ice cream or yogurt container and designate an area to place it in, like the lobby or cafeteria, for example. It's much easier for your employees to bring used batteries to work since they come in every day, than to have them accumulate at home. Once your receptacle is full, have a volunteer take the used batteries to your local recycling depot. You'll probably find that you need a bigger container as you'll see it fills up pretty quickly when even a few people begin contributing.

Choose rechargeable batteries for regularly used devices. Not only does this save you money, but batteries need to be disposed of properly when they're no longer chargeable. This impact can be great in an organization. Think of an organization with many flashlights or electric gadgets, for example, then consider how many batteries they would save on a yearly basis.

54. Toilet Paper and Tissues

When buying toilet paper or tissues, buy the kind made from 100% recycled materials. By doing this, you eliminate the need to cut more trees and help the recycling streams. More and more of these products are now made without any whitening agents like chlorine, which is a good thing.

55. Calendars

Can you imagine how much paper is needed to put all those wall calendars in offices and organizations around the world? Can you now imagine the impact it would have if only half of these organizations bought erasable wall calendars instead? Making this change requires a single investment and is probably one of the easiest tips to apply in this book.

Electricity

56. Sell Your Own Electricity

If a small river flows on your property or you have significant winds blowing in your area, you can invest in renewable energy sources and sell your surplus electricity to a local provider. This not only saves money but can make your organization money, too. If your energy source is intermittent, this method can offset the costs of using network energy when your own system is not enough.

57. Use Alternative Sources of Energy Without Investing

More and more, electricity companies are starting to offer consumers a choice between a main source (which is often coal, unfortunately) and an alternative source like wind, hydro or solar power. In some regions you can choose that the electricity company buy the equivalent of your consumption from a renewable source with the click of a mouse. This will also entitle you to advertise your commitment to the environment and thereby encourage others to adopt the same practices.

58. Solar Power

Solar power takes the energy provided from the sun and converts it into electricity. Solar power is considered by most to be the greenest of the renewable energies. This industry is growing and new ways of using it are being developed, such as using solar shingles made out of solar cells combined with slate, metal or asphalt roofing. This is done by placing solar panels on the roof, on a post, or away a little, preferably facing south for optimum performance and dependent on municipal regulations. Various web sites describe their installations, the different products offered, and the savings associated with each. Solar panels are also easier to find and more cost efficient these days. Most city organizations operate in buildings with flat roofs which are ideal for solar panel placement since they're not visible from the street. Another way to use solar energy is through passive heating. In a nutshell, passive heating works by absorbing as much of the sun's heat as possible, typically by placing more windows on the south side of a building.

59. Use Electricity at the Right Time

Commercial and industrial electrical consumption accounts for 45% of all greenhouse gas emissions. Try to avoid consuming unnecessary electricity during peak hours, which are usually from 7:00 AM to 9:00 AM, and 5:00 PM to 7:00 PM. These are times when most people get ready for the day or return home from school or work. Check your utility bill for surcharges or add-ons so you can see when your business consumes more power. Electricity providers need to plan to support demand at peak times, so try to put measures in place to reduce your electricity requirements during these periods. One way you can do this is to offer your employees the opportunity to come in at staggered times. This can be based on operational needs and personal needs. You may want to go bi-energy and heat with gas at those times and normal electricity the rest of the day.

60. Wind Power

Wind power is effective in areas where sustained winds are present. Wind is the world's fastest growing renewable energy source. It converts kinetic energy into electricity with a turbine. Take time to pick the right location for a wind turbine and the effort will serve you later by being more effective. Also, wind power disrupts wildlife, so make sure that this impact is limited as much as possible.

61. Right Heating and Right Cooling

Right heating is the concept of heating just enough at the right time. Using programmable thermostats or programming you central heating system is how you achieve this. The main advantage is that you can reduce the temperature when not required. For example, when you sleep, you can easily reduce the heat by a few degrees and also when you go away for a vacation, weekend get-away, or even to work. This can easily help you save 5 % to 10% of your energy consumption on a yearly basis, and these devices are better at keeping the desired temperature. Because a programmable thermostat's precision level is higher than that of older models, you're certain that the room stays exactly at the temperature you've set.

Formal dress codes with long sleeve shirts are less suited to summers or warmer climates. Requiring your employees to dress formally also requires cranking up the air conditioning to maintain comfortable conditions. Start by having casual Fridays, where people can wear short-sleeve, open-necked shirts so your organization's temperature can stay a little higher. After that, you can adopt a summer clothing policy that is in effect between specific dates you choose. If you feel this may not be well received by customers, provide closet space so your staff can keep more formal attire for certain occasions.

62. Compact Fluorescent Lights (CFLs)

CFLs are four times more efficient than traditional bulbs and last longer. Although they are contain gas, if disposed of properly at an approved recycling center, the benefits of a CFL's 50% to 80% energy savings more than compensate for the gas used in production. Make sure you put these bulbs in rooms where the lights stay on for a good period of time, since turning these lights on and off reduces their life span. If you turn on CFL-based lights and need to leave the room but will return in just a few minutes, leave the lights on instead of turning them off.

63. Lighting Tips

To make sure you use lighting energy more efficiently, start by having your bulbs and neon lights cleaned frequently so the light generated is crystal clear. Second, luminosity is better if most of the wall colors are pale. Third, when a task requires focused task lighting, turn off surrounding lights when not required.

64. Replace your Light Bulbs

Did you know that traditional bulbs use 5% of the energy needed to light and 95% is emitted as heat? With CFLs, this ratio is 80% light and 20% heat. Don't throw your traditional light bulbs away, but as your bulbs burn out, start replacing them with CFLs to gradually use less energy.

65. Assess Your Energy Opportunities

Do a complete tour of the work area you want to renovate and perform an energy audit. An expert can do this more efficiently, but common sense is enough to start with to take advantage of the most obvious opportunities. Questions to ask yourself are: Is natural sunlight used to its full potential? Is the temperature set correctly and does it go down after work hours? Can CFLs be put in some areas? Is the Energy Star logo on all equipment? Can green power be purchased? Are my electronic devices setup for sleep or hibernate mode?

66. Motion Activated Lighting

Don't just rely on yourself or your employees to remember to turn off lights when leaving a room. Use motion sensors to activate some lights. This type of lighting can be used outside, like at the emergency exit door which can also discourage burglars. Inside, low traffic areas or rooms that are busy at given times such as the break room are also good places to install motion activated lighting. If you want to install these lights but are faced with the fact that sometimes there's no movement and the lights turns off, find a light with a timer attached to the sensor so that if there is no movement for some time, it doesn't turn off.

67. Smart Strips

Smart strips are a great way to reduce energy not in use. Most appliances use energy even when turned off. This can be to maintain an internal clock or simply to power that little light that tell you the device is turned off. In any case, we can all agree that this energy is wasted. Smart power strips are designed so that you plug in a main appliance and peripherals. Once the main appliance is off, all the power going to the peripherals is also shut down. Consider smart strips for your PC, stereo or TV system.

Water

Did you know that unsalted water represents only 2.5 % of the earth's water? Did you also know that two-thirds of that water is composed of glaciers and eternal snow? Lakes and rivers represent only 0.3% of all the unsalted water present on the planet. So, if someone you know is not yet convinced that we need to put effort into preserving our water you can use these arguments. You can close your discussion by explaining that we can go several weeks without eating, but will die after a few days if we don't drink.

68. Water Conserving Toilets

About one-fourth of the average home's water consumption goes out the toilet if water efficient toilets aren't installed. In the workplace, water consumption related to toilets is not as important but is still significant. New water conserving toilet models use less than half the water of traditional toilets (1.5 vs. 3.6 gallons), on average. Installing water-friendly toilets saves water system capacity upstream and downstream. Even if you're not ready to buy a two-flush toilet, you can still improve your water consumption by installing a newer model.

Showerheads, faucets, urinals, and toilets sold in the U.S. must now achieve a maximum flow rate in terms of water used. Select ones that exceed federal standards to ensure you promote higher standards. You can find independently verified products by consulting the Environmental Protection Agency's WaterSense program.

If you need to shop for a new toilet for your organization, choose a two-flush toilet. These will save you 50% of the water needed to flush a normal toilet when you're only flushing liquids.

69. Cold Drinking Water

Your employees who like to drink cold water often let the water run for a minute or two with the tap open to keep from drinking room temperature water. If your organization has a mini kitchen or a refrigerator that employees use for their lunches, keep a container filled with water in the office refrigerator. This is an opportunity to avoid wasting all that water. You'll save gallons every year.

70. Recuperate Gray Water

You can install a system that recuperates gray water from your office's sinks and washing machines. A gray water system filters this waste water so you can reuse it afterwards for outside watering or your toilets. When combined with rain water, gray water can be sufficient for all the toilets in your organization.

71. Responsible Water Consumption

Water is needed for a variety of tasks like cleaning (clothes, dishes, people, cars), drinking, heating, garden work, toilets, etc. While carrying out these activities, make sure not to waste any water. Public water is treated and therefore needs processing with chemicals, filtration, and so on. As a taxpayer, you pay for all of this. Also, although water that's returned to the environment is treated according to regulations, it's not the same as it was before going into the water system, therefore it can potentially affect bio-diversity downstream. Your manufacturing processes may need water, too. Make sure water is used responsibly and that alternate solutions are looked into to avoid wasting water.

72. Water Barrels

Place water barrels outside to collect rain water from your workplace's roof. In most climates, there's enough rain that your barrels will hold enough water to keep your flower beds and small gardens well watered. Have your lawn cutting company use this reclaimed water before using fresh water from the tap. You can even put your barrel on an elevated platform or up on cement blocks so you can plug in a hose for watering. If you look at the marketplace, you can find barrels already modified to plug in a hose.

73. Storm Water Runoff

Most commercial operations have overly large parking lots and this increases storm water runoff. When a storm occurs, rainwater falls on the asphalt and picks up pollutants that are then evacuated into designated overflow areas. Since asphalt doesn't really absorb water, pollutants are much more concentrated within this water. When planning you site, keep this in mind and try to minimize the potential problem. Also consider the fact that heavy asphalt concentration increases surrounding temperatures and that asphalt is made from petroleum. The less asphalt you have, the greener you are.

74. Water Early in the Morning

Many companies have automatic sprinkler systems or have their lawns watered. If this is true in the case of your organization, make sure watering is done early in the morning. When you water in the morning, the water has time to be absorbed before it gets too hot. Evening watering is better than midday, but although absorption is good, the ground stays humid for longer, a condition in which fungus thrives!

75. Leftover Water

You have employees that have their own water bottles. When they come home from work with a little bit of water left in their water bottle, encourage them to use that water for watering indoor or outdoor plants rather than wasting the water down the drain. Plants aren't picky about their water source!

76. Infrastructure: Water and Power

As the population grows and becomes more concentrated in and around cities, the need to have bigger water treatment plants and power systems usually arises. By reducing your organization's consumption of power and water, you help postpone and maybe avoid this increased need. Alternatives to building new water treatment facilities are now beginning to surface, such as using non-treated water from lakes, rain or rivers, and directly using this water for watering lawns, flushing toilets, and so on.

77. Cooler External Walls

If you live in the northern hemisphere and one of your workplace's external walls faces east, west or south, that wall is probably very hot when the sun shines on it in during the summer. This heat accumulates throughout the day and will often be hot to the touch several hours after sunset. This contributes to global warming. To offset this, plant shrubs and trees close by if space allows. When space is limited—especially in urban areas—you can plant vines that climb and stick to the wall. These vines take up almost no room on the ground, need no maintenance, offer good shade for birds, and absorb the heat, making the inside of the building cooler.

78. Composting

Doing your own compost at work can provide several advantages. The first advantage is the actual use of the compost. As a landowner, you can use it yourself, but it can also be used by employees who contribute. Second, composting requires very limited space for most organizations, while the space required for an entire city's organic garbage is quite considerable and is becoming quite rare. City garbage dumps are generally overused and these sites are more remote than they used to be. Third, you cut down on garbage disposal costs and its effects on the environment. As you know, having less garbage on each property means fewer garbage trucks are needed (less gas), requires fewer tractors to manage the dump, and so on. An added bonus is a leaner municipal tax bill. Some cities have a compost program where they pick up organic waste from houses and industries, and tax payers can then go get compost once it's ready. This is a good solution, especially when people don't have room for or can't compost for some reason, but the second and third benefits of composting listed above are not achieved that way.

There are tricks to effective composting and one of the most important is maintaining the ratio of dry ingredients (66%) to wet (33%). Good examples of dry ingredients are hay, dried lawn clippings, and dried leaves. On the other hand, leftover table vegetables and garden vegetables are good examples of wet ingredients. Be sure not to put meat or any animal products in your compost, as these will attract unwanted visitors to your compost pile. If you have any doubts about what can and can't be put in a compost pile, check out: www.gardenorganic.org. At work, find a volunteer to be in charge of the compost pile, or set up a rotation amongst contributors. If your organization has a cafeteria, include the cafeteria in this project. You'll be surprised to see how much organic waste material is produced by a cafeteria on any given day.

79. Mulch All You Can

Leaving leaves on your lawn not only saves you the time you would've spent picking them up, but also keeps the lawn healthy. Cut leaves that are composted turn into nutrients for you soil, which then saves time and money because you avoid the use of commercial fertilisers and compost. Using a lawnmower that turns leaves into mulch increases the speed at which this material is incorporated in the soil.

Cover flower beds and trees with 3" of organic material (mulch). This conserves water, adds humus and nutrients, and discourages weeds. It also gives your beds a nice, finished appearance.

80. Trees and Shrubs on the Property

Making sure your workplace has plenty of trees surrounding the building helps keep the air surrounding the walls cooler by providing shade. This also contributes to keeping the air clean, something which is very important in urban areas.

81. Proper Tree Positioning

If you live in a northern country, plant evergreens on the north side of your organization to provide protection from winds. Plant trees with leaves on the southern side for shade in the summer and to let in additional sun during the winter.

82. Grass Selections

When preparing your yard for grass, select the optimum type of grass for your area. Most popular grass seeds require more maintenance. Choose one that doesn't. Clover is an example of a type that's usually low maintenance and requires very little water to remain green.

83 Keep the Grass Long

We like golf course looking short grass but this type produces a lot of humidity and therefore requires more watering to ensure it doesn't dry out. By keeping your grass longer, moisture is retained and unwanted weed seeds can't grow as easily because the environment is too dark.

84. Perforated Hose

If your lawn and plants require watering and you have employees perform this task, have them use a perforated water hose. With a perforated hose, water is better absorbed by the plants instead of the ground becoming flooded.

85. Use Native Plants

Choose native plants when selecting plants for landscaping. Native plants are accustomed to your climate and pests, and therefore need less human intervention to grow and blossom. Choosing native landscaping also reduces the need to water.

Indoors

86. Recycle Your Carpet

When you buy new carpet for your organisation, have the installers take your old carpet back for recycling. If your provider refuses to do this, don't be shy about shopping elsewhere to find one that will. Stores need to adjust, and you as the customer need to let your expectations be known.

87. Leaking Water Taps

Fix leaking taps, even if the leak seems small. You'd be surprised how much water can be saved this way.

88. Cleaning Products

Did you know that almost every stain commonly encountered can be removed with vinegar and/or baking soda? Both these products are cheaper than typical commercial products, so why buy 35 different products to do the same job? This tip may not apply to industrial processes, but for an office environment it surely applies. Use a cleaning contractor committed to clean and green products.

Green cleaning products are safer for workers since traditional products often contain large numbers of hazardous chemicals. These products release toxins that can affect health and productivity in your organization. Authenticity of green products can be verified on Green Seal's website at: http://www.greenseal.org.

89. Lights on a Timer

Keeping outside lights on while everyone sleeps is nonsense. If you really need to have outside lights, make sure they're kept to the minimum and run off a timer.

90. Water Heater Insulation

You can buy an insulation blanket designed to fit on your water heater. This is not an expensive product, and will save you money. If your organisation is large and you have an industrial water heater, talk to your plumber. He probably knows where you can have a custom insulation blanket made.

91. Pipe Insulation

Insulate your water pipes with foam. This will decrease heat loss between your water heater and sinks, showers, and appliances.

Events and Seminars – Offsite

92. Ecological Procurement

Any purchasing activity should be done with the environment in mind. You should ask potential suppliers about the environmental impact of their products. Be critical of a product's environmental value. Make sure a potential supplier's products comply with official standards or norms such as Green Seal and Scientific Certification Systems. These standards can be used to impose specific ecological demands upon your suppliers. Unfortunately, many goods don't fall under the umbrella of these specifications. When buying these products, ask sellers for information on the ecological characteristics of a product being considered and compare these products with those offered by the competition to evaluate who offers the most environmentally friendly products. In most cases, this evaluation is largely subjective. In an ideal world, all the environmental impacts of a product from cradle to grave would be taken into account.

93. Sponsor Selection

Sponsors should be evaluated and selected with regard to their environmental practices and policies. Even if sponsors give away their products or services, you should favour ones that are respectful of the environment. Does your sponsor distribute documentation made from recycled paper or offer reusable mugs instead of paper cups? Asking for products made with respect to the environment leads sponsors to begin offering such products.

94. Food Services Selection

Food and liquid consumption has the potential to generate a great deal of paper, plastic or other waste. Event planning offers a great opportunity to educate participants on reducing their food-related waste generation. There are many ways to conserve. Use real plates, glasses, utensils and cups instead of paper. Offer tap water at tables instead of bottled. Ask your sponsor to distribute mugs at the start of a conference and offer an incentive to participants for reusing their mug.

95. Choice of Location

Select a central location close to the airport. Choosing a downtown location close to the metro, a bus stop, or a train station will make your event more accessible from public transportation.

96. Communications: Ecological Promoting

While promoting your event or contacting participants, you'll have multiple opportunities to communicate your environmental beliefs and encourage others to make an effort toward going green. Give your communication strategy serious reflection so your messages support your environmental beliefs.

97. Safety

Good communication between the safety team and the people in charge of an event will lead to uncovering any safety policies that may have a negative impact on the environment and finding solutions. Identification tags can be made with recycled paper or plastic. The same goes for access cards. If cars are part of the safety program for your event, why not use hybrid cars, or at least have people turn off their engines when not moving.

98. Participant Lodging

If event participants are arriving from out of town, evaluate the ecological elements relevant to their stay. First, consider location. Choose a central location near the event and major modes of public transportation. Organising shuttle transportation can be another way to save on gas. Why not give away public transportation passes to attendees for the duration of the event? Or why not hold your event at the hotel if most participants stay at that hotel? This has several advantages. You won't need to go out to eat, no one will be late because of traffic, and so on.

Travel and Transportation

The Green Globe program measures performance for the travel and tourism industry. It measures greenhouse gas emissions, waste water management, noise control, air quality protection, energy efficiency, fresh water use, solid waste minimization, land use, and ecosystem and local impact. For more information you can go to: http://www.greenglobe.org.

Policies

99. General Travel Policies

Here are some policies you can put in place to make a difference while traveling:

- Ask that mass transit be the prime transportation mode;

- Use fuel efficient rentals if absolutely necessary; and

- Book accommodations in hotels that are members of the Green Hotel Association.

100. Carpooling

Encourage your employees to carpool. The benefits of carpooling are easy to understand, but what's difficult is actually doing it. People like to have the luxury of coming and going as they please. Instead of quitting before starting, here's a little recipe you can tell your employees. First, find someone close to your home and try carpooling to work just once. Don't commit for a month or even a week, just do it once. Chances are that you'll find it's not that bad at all. Now, try to do it once a week. This way you'll still have the flexibility to come and go four days a week. Chances are that you'll find that's not bad either. Once you're successful at this, try increasing the frequency or the number of people in the carpool. The idea is that in order to be effective, car pooling needs to be adaptable to the realities of the car poolers. When aiming for long-term success, you must find a comfort zone that suits your life and is sustainable.

101. Premium Parking

A good incentive you can offer your employees to carpool is to have designated parking spots for carpoolers located closer to the entrance. There's no real cost involved, but there are advantages for the people that make that effort. As you can probably imagine, this measure works better in bigger companies where parking lots are huge, but don't be fooled. Even in small organizations, this incentive can be successful.

102. Setting the Example in Transportation at Work

If you're responsible for buying vehicles for your company, buy energy efficient models. Encourage people to work from home, therefore reducing the company's overall indirect ecological footprint. If you work on the road, practice good driving habits to reduce your gas consumption.

103. Work From Home and Flex Time

Technologies like instant messaging and video conferencing are now available to make effective telecommuting a reality. Encourage your employees to do so by organizing the work schedule accordingly. Telecommuting will not only save them time and gas, but will save your workplace energy. It can even save you real estate if you have a significant workforce that can work from home. You will need to have a performance-based culture and not the more common attendance-based culture, and not every organization is ready for that. Allow flex time so your employees can stay longer when the traffic outside is terrible, or if they can't get to work in the morning.

Also, consider the possibility of organizing your office for four, ten-hour days instead of five eight-hour days. This is known as a consolidated work week. This will reduce the energy and time spent on commuting by 20% and give you some lovely three-day weekends. Your organization will also benefit from having employees who are less stressed.

104. Hybrid Cars

If the size fits your needs, buy a hybrid car. For most models, the hybrid version offers more power than the gas version while consuming less. It used to be true that the only well-known hybrid was the Toyota Prius, which was and still is highly recognizable. But not everyone likes the particular design, and hybrid models are now available in traditional designs that don't stand out. Buying a hybrid will save you 20% to 35% on gas. This may not be enough to compensate financially for the higher price tag if you don't rack up a great deal of driving miles. If that's the case, maybe the bus, train, or carpooling are better solutions for you. Don't buy hybrids because you feel they're cool. You may also want to combine buying hybrid with other green best practices like keeping your car longer to make it worth while.

There is a simple exercise you can do. Calculate how much mileage your organization's car or fleet has traveled. Divide this by the average miles per gallons an average hybrid car gets and multiply this by the cost of one gallon of gas at the pump today. Once you have that total, compare it to your actual spending on gas and see how much you would have saved.

You can also use that number to calculate how much time it will take you to offset the potentially higher initial cost for buying a hybrid.

105. Electric Vehicles

Electric vehicles aren't dead. Although the dream of having 100% electric cars like in the sci-fi movies is not there yet, many applications are now possible. Golf carts are probably the most widely used for individual traveling, but now there are electric scooters, buses and bikes.

106. Gas Emissions

Consumers focus on gas consumption because of the impact on their pocketbooks, but their real concern should be gas emissions. The less gas a car uses, the less it pollutes. Make fuel efficiency an important element in decision-making when buying a vehicle for your organization. Also, position your location so it's central to where your product or service is delivered. If you rely on distribution centers as part of your distribution strategy, place your goods closer to where they will be sold or used.

107. Keep Your Car

With all the over-consumption and focus on image, people have become used to changing cars every two to three years. Did you know that good cars made by companies like Toyota, Honda and Subaru can go 10 to 12 years without rust and major mechanical problems when properly maintained? Of course, over time, models will change and your car might look out of date, but what's the point of owning a car? Style or taking you from point A to B safely?

108. Scooters

If your employees can travel light (weather permitting) why not use a scooter—preferably an electric scooter? Obviously scooters burn less gas than cars, but they also bun less fuel than motorcycles. For short commutes scooters are a great alternative to bikes. Bigger scooters can carry two people so this may be an option, too. Scooters are not considered green by many people because their engines are not efficient. In fact, they're less efficient than a car engine. Our point of view is that, although less efficient, the scooter is better suited for individual travel because it's much smaller than a car. Going to work is usually done solo, so there's less gas consumed per person with a scooter. Do you really need a four passenger car to commute to work alone?

109. Engine Oil

Buy oil that saves energy. Using oil with a lower grade of viscosity than the manufacturer's suggestion can improve gas consumption, especially in cold weather.

110 Unload Unnecessary Weight

Ensure your organization's vehicles don't carry unnecessary materials on a daily basis. Obviously you don't want to get rid of your emergency kit or spare tire, but do you really need that toolbox that's only been used once? By reducing non-needed cargo weight you will burn less gas.

111. Avoid Ethanol

Ethanol is marketed as being green and as the 'right' alternative to petroleum. Although at first even scientists held this opinion, the wind has now shifted as we see what the real impact of this alternative fuel is. Biodiesel is made from renewable resources such as corn and then transformed into alcohol. The problem is that now that plants producing ethanol are becoming bigger to produce more for the market, they consume more natural resources which normally would and should be used to feed humans and animals. With all the people in the world suffering from hunger, can you imagine if, overnight, we all turned to ethanol? Moving to ethanol avoids addressing the real problem with cars today, which is that we simply drive too much and do so at the expensive of the environment.

112. Cruise Control

Have your employees use cruise control when possible while driving on the highway. It's been proven that cruise control saves gas. It's also going to save you dollars by preventing speeding tickets.

113. Smooth Driving

Driving at an even speed and not exceeding the speed limit will reduce the emissions produced by your car's engine. In this fast-paced world, it's easy to get carried away and give in driving fast and darting in and out of traffic, but resisting the urge will reduce your carbon footprint and the amount you shell out in speeding tickets.

114. Speed Control

Every time you go over 60 miles per hour or approximately 95 kilometres per hour, you car engine becomes less efficient. Even if you're tempted to go over the speed limit, take a deep breath and stay in control!

115. Tire Pressure

Check car tire pressure regularly. Low pressure results in less efficient tires by increasing resistance and decreasing fuel efficiency. Be sure to double check tire pressure when temperatures change rapidly.

116. Oil and Oil Filters

If your organization has a fleet of vehicles and is equipped to do its own oil changes, great. But don't forget to have your filters and oil recycled. If you have a small fleet, take your used oil and filters to a local auto parts store which will probably offer to recycle these items for you at no charge.

117. Brakes and Alignment

While driving straight, hold your steering wheel very loosely—almost not touching it. If it tends to shift to the right or left, have your front-end alignment checked. Bad alignment makes your car burn more gas.

118. Ventilation in the Car

When on the highway or going fast, favour your car's ventilation as opposed to its air conditioning system. Use the air conditioning system only when absolutely necessary. Auto air conditioning requires about the same amount of gas as driving with your windows down. This last option is the least desirable since it also makes your engine work harder.

119. Old Car Disposal

When you need to get rid of an old vehicle, do it the right way. Check to see if there is an official program for this in your area. Before your car is left to rest in these programs, all fluids and dangerous products (CFCs, mercury, oil, gas, etc.) are removed. Batteries and tires are recycled, metal from the chassis, too. Sometimes these programs also offer financial incentives, like new car rebates, public transportation discounts, and so on.

120. Roof Racks

If your vehicles are equipped with roof racks, don't leave them on permanently. Use them only when required. These accessories make a vehicle less aerodynamic, making it less efficient from a gas consumption stand point.

Travel

121. Carbon Offset

Find ways to avoid any air travel that isn't absolutely necessary. Taking a plane is, by far, the least eco-friendly way to travel. If you must fly, at least do the right thing by buying carbon offsets. Carbon offset funds contribute to projects with ecological value, such as planting trees or capturing gas, like: Carbonfund.org.

122. Hotel Linens and Towels

Use hotel linens and towels more than once. Some hotels do this by default and some will give you the option. Hotels use a lot of water every day, and face it, when you're at home, do you wash your sheets and towels every day?

123. Transportation for Conferences

When planning work conferences or large gatherings, keep in mind that people needing air transportation will be the biggest contributors to green gas emissions. You can reduce this by holding your event in an area where direct flights are possible. You save this way because taking off and landing require more gas than just flying at a constant speed. Through your gatherings, you can also promote car pooling and public transportation by adapting the schedule to fit with local bus stop times, or giving gifts and discounts to carpoolers, for example. You can also organise busses if people will need to move from the hotel to your conference, thereby avoiding the need for rental cars.

124. Traveling for Work

When putting travel policies in place for your workforce, consider the train. In certain areas this may be almost as quick as other travel methods. Also, post a few questions people should ask themselves before actually going on a business trip. Can three trips be combined into one or two? Can a meeting be replaced by a teleconference? Can a colleague who will be traveling in that area represent you?

125. Electric Systems

If you work on the road, or if your organization has a few vehicles or a fleet, as a standard policy, have all accessories turned off when the vehicle is parked. This will produce less strain on the battery when the engine starts and therefore your engine will have less trouble starting back up.

Construction

126. Choosing Where to Locate your Organization

Does the city you're about to locate in have plenty of green spaces, parks, trails, recycling and composting programs? This is a good indicator of a city's commitments to green living. Also, when you know which city you're going to set up your organization in, find a location that will minimize employee commutes. Do what's called a spaghetti chart by taking a map and drawing all the transportation activities that will happen on a normal work day. After that, calculate the distance traveled. Then try to position your organization elsewhere on the map and repeat the exercise. Make sure individual travel is accounted for, such as workers getting to work or products being shipped to a customer, and exclude mass transportation such as public trains. This activity is very helpful in selecting the best location.

127. LEED Certification

LEED stands for Leadership in Energy and Environmental Design. This certification was established for governmental buildings, but is now used in the private and commercial sectors. Having this certification tells your community and customers that you care about the environment.

Furniture and Appliances

128. Furniture Selection

Favour office furniture made with natural fabrics such as cotton and ramie. Traditional office furnishings can increase office air pollution through the emission of volatile compounds from their glues and finishes. Buy furniture that can be easily dismantled at the end of its life. If wood is used in the furniture you're looking for, try to find furniture made with recycled wood or wood from FSC forests. Another way to preserve forests is to buy second hand. Small bookshelves, low enough for a child to reach, can often be found at flea markets for very little money.

129. Recycle Your Refrigerator

At some point you may have to dispose of an old refrigerator. Some electric companies will offer to recycle your appliances, and in Europe, retailers are responsible for taking back your old appliances.

130. Energy Star Products

There was a time when the Energy Star logo was only on a few products such a washers and driers, freezers, and so on. Now, this label can be awarded to more than 35 types of products. Energy Star is an international standard for energy-efficient electronic equipment established by the U.S. Environmental Protection Agency (www.energystar.gov). More often than not, more efficient products are more expensive than less efficient models because their technology is different or more refined. Before selecting your appliance, make sure you account for all the costs you will incur during the life of the product. Consider the selling price, but also the energy, water, and maintenance costs, year after year. Energy Star features on machines like computers are not automatically enabled, so make sure you activate them after purchase.

131. Consider How and Where Materials are Produced

When choosing construction materials, consider how the producer made the product. Was it manufactured with recycled materials? Was the manufacturing process respectful of the environment and done with sustainable methods? Did the product travel halfway around the world or was it made just a few miles away?

132. Heating and Cooling

When selecting a system to heat or cool your organization, take into account a few of these elements: First, select the most energy efficient appliances with the Energy Star logo. Also, make sure you buy the right sized system to fit your needs. Don't go overboard and install enough heating or cooling to climate control your whole neighbourhood. Prioritize heating from hydroelectric power over burning gas, for example. Once installed, make sure you follow instructions for system maintenance so that its remains optimally effective.

133. Solar Water Heater

In areas where it's sunny enough, solar water heaters are a good alternative to traditional electric or gas water heaters.

Structures

134. Choosing to Renovate or to Build

As your organization grows and evolves, there may come a time when your existing structure is no longer suited to your operational needs. Before deciding whether to build from scratch or renovate, you need to reflect on a few things more deeply.

First, remodelling does not require as many materials as building a new structure. Remodelling can also be done in smaller phases (and at a lower cost) so it can become part of your green plan over a few months or years, depending on the overall scope of the renovation. Progressing is smaller phases also lets you involve more people or employees in your project and results in a smaller carbon footprint than new construction.

If you do decide to build from scratch, you need to understand that this will take a lot of time and human as well as financial resources. There are many things to consider, such as interruption of your organization's activities, the chaos of construction, and all the decisions that will need to be made regarding materials, site selection, equipment, and so forth.

When planning to build a green building, overall life cycle needs to be considered. This often means that initial costs will be higher, but maintenance and energy cost savings will offset this initial expense more quickly than you think. Building a green building will also contribute to better quality air and a reduced carbon footprint. Remember that triple bottom line—People: Planet: Profit—discussed in the beginning of this book.

A big part of the decision on whether it makes the most sense to build or renovate will be based on your long-term strategy. First, what will your organization's growth be like in coming years? If the answer is that you're growing very rapidly, this almost certainly excludes renovation. If you're unsure about your growth, it may be worth investing some time in exploring as this can affect your building plan, but more importantly, your organization's overall future. Also, think about your workforce's location. In your growth plan, do you intend to have people work from home or on alternative shifts?

Your existing building's current state and performance is also something to look at. Whether structural, energy loss related, or environmentally related, if your current building faces several green challenges, it may be better to go with new construction. Correcting existing problems can be very difficult and costly.

Do you plan to sell or close your organization in a few years? Some non-profit organizations are designed to address specific situations such as a natural disasters or recurring seasonal issues. In those cases, a new building will probably incur more costs that will not be recuperated due to the short time your organization will operate.

Are there any grants for green building applicable to your situation? More and more countries and regions have implemented programs to support sustainable building features.

Finally, if you do decide to move out and build a new building, make sure you dispose of the existing structure responsibly.

135. Recycled Materials

When building, try to select materials made from recycled products. More and more options are offered in that category. Just ask your local hardware store.

136. Windows, Doors and Ventilation

Whether your organization is located in a very cold place or in a warmer climate with only a few cooler months a year, ensure your windows don't let cold air in or heat out. Drafty windows are common in older construction that's not well insulated to start with, but are also common in newer buildings where insulation is not as effective as it used to be. There are a variety of products available at your local hardware store to improve insulation. To check if you have air infiltration, bring a lit candle close to the window. If the flame is moving you need to insulate. When choosing new windows and doors, invest in energy efficient models. One other thing to consider: Windows that open using levers are more effective in the long run because they seal better. Sliding windows eventually lose their tightness.

Ventilation can account for 20% of a commercial building's greenhouse gas emissions. In most of these buildings, windows can't be opened and air is pushed around by machines, often resulting in poor air quality if filters and devices are not maintained properly. The best solution is to use natural ventilation.

137. Carpet

If you must have carpet, keep the following best practices in mind. First, carpets need to be cleaned very regularly. Second, select carpet made with 100% recycled materials. This saves the resources and helps landfills. Remember that rugs are treated with stain-proofing chemicals, moth-proofing pesticides, and more. Wall-to-wall carpeting can't be taken up for a good thorough wash, and pollutants can settle deeper than vacuums can reach. Dust mites are a common trigger for asthma, the rates of which have doubled in children since 1980. Pesticides and herbicides sprayed on lawns, lead dust from your neighbours' renovation project, and practically anything blown or tracked into your offices can settle into carpet for years.

138. Stone Tiles

Like ceramic tiles, stone tiles are very durable. One thing to look at, though, is the origin of the stone. The best option is to buy locally. Because stone is heavy, the environmental impact of transporting it over many miles is magnified.

139. Ceramic Tiles

Ceramic tile is very durable and the resources needed to produce it are not over-exploited. Also, more and more ceramic tiles are made with recycled materials like glass and clay. Try to favour recycled materials. Ceramic also requires less maintenance since it's easier to clean than carpet.

140. Green Roof

In urban areas where the roofs are flat, why not put up a green roof? A green roof is a roof that is covered with a growing medium. Not only does this help reduce global warming, but you can also use is to grow vegetables for your employees and plants you like, such as flowers that you can cut to bring inside. A green roof also reduces heating and cooling needs because it offers better insulation than a typical roof and is good at rain water management. Your roofing has a strong impact on your building's efficiency. For example, a dark roof absorbs and emits more heat than a light roof, and therefore increases cooling costs.

141. Green Renovation

Always try to reuse the material being removed when renovating. If you can't use it yourself, donate it to your local recycling center. Often times they will accept construction material. Also, even though the building you're renovating may be old, take the opportunity to put in better insulation, low water consumption toilets, etc.

142. Construction Materials

When select material for your construction, select those made from renewable resources and that are good for air quality or the well being of people that are going to live in this new building. There are many framing choices, such as steel, which uses a lot of energy to make but is easily recyclable. If you use wood, on the other hand, make sure it's certified by the Forest Stewardship Council (FSC).

Insulation is the most important thing when looking at energy use in a building. Many green options are now available if you go to your local hardware store and check out formaldehyde-free and recycled content insulation.

For the walls, why not go with Structured Insulated Panels (SIPs)? These are very durable and arrive at the construction site pre-cut. Another option is recycled gypsum.

143. Paint Type

If you need a synthetic paint, opt for water-based latex paints over oil-based alkyd paints. Water-based latex paints have lower volatile organic compound levels. Natural paints are usually made from citrus and other plant ingredients, like milk protein or clay. They're also free of preservatives and biocides. Natural paints are best suited for drier areas, as they're less resistant to mildew and mold.

Milk paints are virtually odourless and are made using the milk protein casein and lime. They contain no solvents, preservatives, or biocides, though some do have synthetic ingredients like acrylic and vinyl. Milk paints come in powdered form. Once opened or mixed with water, they should be used quickly, as they can mold if left to stand for a few weeks.

Whitewashes, which only come in white, contain only lime paste, water and salt. They are a low cost alternative that, like milk paints, are more fragile and are best applied to plaster, cement or stucco walls. Check the back of the paint can for VOC levels (you can also look online for the "Manufacturer's Safety Data Sheet"). An ideal paint has fewer than 150 grams per litre. These are often labelled "low-VOC" or "no-VOC."

Before buying paint, check to see if you have leftover paint from previous projects that can be used. If you find paint that's no longer good have if recycled. If you must buy new paint, buy only as much as you'll need to complete the project. You can still go get more if need be. Some stores may let you return excess paint, but if that's not possible, store it safely or dispose of it in accordance with local municipal regulations.

Finally, donate unused paint to a local organization that needs help. Even better, organise a team-building activity to paint your local youth or senior citizen center.

144. Painting Tips

When painting indoors, open all windows and use fans to vent fumes. Pregnant women and people with allergies or asthma should not paint and should stay out of a freshly-painted area for at least 48 hours. Try to have painting done over a weekend or at night so that most of the fumes are gone when employees come back to work.

When sanding or removing paint, wear a dust mask or respirator and keep the area well ventilated, since this process generates carcinogenic crystalline silica dust. In pre-1978 constructions, test painted surfaces for lead before sanding. If lead is found, contact a professional for remediation.

Tips for Employees

Although people buying this book are very interested in reducing their workplace's environmental footprint, not everyone owns an organization or has the power to make some of these changes. This next section was specifically added with these people in mind, so that employees can change some of their daily habits at work and at home.

1. Bring Your Lunch

Not only are pre-packaged lunches a good example of over-packaging, but the quality of the food inside is highly debatable. Bringing your own lunch is good for you and for the environment.

2. Dry Cleaning

The dry cleaning process uses many chemicals, such as tetrachloroethylene. If this word is unfamiliar, tetrachloroethylene is a chemical that is associated with possible cancers, that can aggravate asthma and allergies, and obviously, is no friend of the environment. A solution is to buy clothes that don't require dry cleaning. If you must dry clean or have a number of non-replaceable 'dry clean only' items in your wardrobe, try removing stains with cold water or by spot cleaning. When selecting a dry cleaning shop, choose one with green policies—reusing hangers and garment bags, for example.

3. Eat at Green Restaurants

It's a common practice in the business world to take important customers out to dinner. If this is something you do, support restaurants with green policies. Choose restaurants that incorporate green practices into their day to day activities, such as by installing low water consumption toilets or featuring organic food on their menus. Other good practices for restaurants are composting, and recycling bottles and plastic.

4. Doggy Bags

It's very common in the United States for restaurants to prepare take-home leftover bags for people who don't finish their entire meal. Most of the time, these leftovers are packaged in Styrofoam because it's cheap, sturdy, and well-insulated. Next time you go to your favourite restaurant, why not bring along your own carry-out container made of plastic?

5. Thawing Meat

In this fast-paced world, it's now common to decide what to cook at the last minute and thaw frozen food on the counter or in the microwave. Using the microwave or stove obviously means using power, which we don't want. Even thawing food on the counter needs heat, often generated by your house's heating system. On top of this, meat is not something you want to leave at room temperature too long as bacteria may develop. The best way to thaw food is to place it in the refrigerator the day before you intend to cook it. This way you reduce the power needed by your refrigerator, since the meal acts like an ice pack in a cooler. Also, as mentioned above, it's better to thaw meat in a cold place to avoid bacterial proliferation.

6. Food Waste

Let's face it, we have more than what we can eat in North America. When planning meals, start with the ingredients in your refrigerator, not with what you'd like to eat. Also, use leftovers for lunch at school or work. Eating only what you need to and not wasting the rest can help feed the rest of the world.

7. Storage Containers

When storing leftovers, try to use glass or even porcelain containers instead of plastic. This avoids possible chemical transfer from plastic containers, such as phthalates.

8. Bring Your Bags

Bring your own bags when grocery shopping. If the product is small enough, decline a bag. Plastic bags take forever to disappear from landfills and the environment. They also cause damage to wildlife. For example, sea turtles in the ocean often die when they mistake these bags for jellyfish and eat them.

9. Eco-tourism

Have you ever tried eco-tourism? This type of travel will bring you to places less crowded with tourists and where you can actually see and learn how people live in a given region or country. As an added benefit, you'll also encourage more regional tourism.

10. Pack Light

Not only will packing light help you when you travel and need to carry your things up the stairs or on the street, but you'll also save the transportation fuel (plane, train and automobile) needed to transport the weight of your belongings.

11. ATM Receipts

When going to the ATM, ask yourself why you request a receipt. When ATM banking first began, there was an understandable culture change away from going to the bank and speaking to a person to retrieve money from your bank account. People were worried that the 'machine' would make errors, or they wanted to be able to validate that, at the end of the month, everything was OK. But now, who does that sort of reconciliation? ATMs are now integrated into our lives and requesting a receipt is a waste of paper and time.

12. Paperless Bank Statements

Ask to receive your monthly bank statements electronically and save them on your computer. You'll save paper and space. Don't forget to make back-up copies of your hard drive every six months.

13. Electronic Tax Refunds

If you're expecting an income tax refund, request that your refund be direct-deposited into your bank account. This option will save paper, money, and time.

14. Make Donations

If you really don't need certain things such as clothes or furniture, see if you can't donate these items to a local community organization rather than taking the easy route and throwing them into the trash.

15. Hemp Clothing

Hemp is a plant that doesn't require pesticides to be grown productively, and its fibre is a very good material for making clothing. While hemp is more famous for its smoking-related properties, clothing-use hemp does not contain the THC that causes hallucinogenic effects.

16. Recycle All the Way

There are three important steps to recycling. First, you need to identify recyclable materials and store them for recycling. Think beyond what you'd usually recycle and go out of your way to gather materials that normally wouldn't be recycled, for example, pick up things left on the street and recycle obsolete clothing or furniture. Second, you have to get your recyclables out to the street for the weekly collection, where available, or take them to your local recycling center, especially if they're over-sized, like pots and pans, or mattresses, etc. The third step is that you have to encourage the recycling industry by buying recycled materials

17. Become a Vegetarian

You probably know that meat requires more energy to make it to your plate, more square footage to produce, and consumes more resources. Sure, changing to a 100% meat-free diet is probably going to be hard for you if you're used to eating meat at every meal, but it's taking the small steps that make a big difference, especially if you're making changes as a family. Try replacing only one meal per week. See how your family adapts, then try to decrease the quantity of meat you buy every week. By taking this slow approach and sticking to it, you will adapt more easily.

18. Rent Movies at Home

Instead of going out to see a movie or renting one, download one from home. Cable companies offer this service now, just ask for it. Not only will this save you gas, but you'll be sure that the movie you want is available, which is not the case when you go to your local rental store.

19. Plant a Small Garden

Planting a vegetable garden brings lots of benefits. First, you grow (organic) plants helpful to the overall ecosystem. Second, you eat good, healthy food that didn't need fuel to transport. Third, you spend time outside caring for your garden instead of staying inside. The physical activity and fresh air are good for you.

20. Give Products a Second Life

When building an outside shed for storing your lawnmower and patio furniture, why not put in a used door or used window? When you need furniture or lamps, why not look for antiques and check yard sales? You'll sometimes be surprised to find these items in good shape at bargain prices. Giving discarded products a second life is an application of one of the big three "R"'s of green living: Reusing.

21. Build and Own Smaller Houses

Small is beautiful, as they say. Over the past two decades, the size of the average house has grown more than 20 %. This extra square footage is unnecessary in terms of excess energy consumption, furniture, gadgets, and gizmos. Make sure you build the right size home and if, eventually, you think you'll need more rooms, renovate when you really need to. You can plan accordingly and design your house with these new needs in mind in order to save costs when you actually add a room or section. Also think of later years when the kids leave the house. Do you want to continue living in that house, and if so, think about all these empty rooms and the energy needed to maintain and keep those rooms clean. You may want to move to a smaller house at that point, too, closer to your children and grandchildren.

22. Avoid Using Pesticides

Pesticides are made from chemicals, and while effective at eliminating pests, they're also destructive to other natural elements surrounding the sprayed area. If fact, it's alarming to know that the long-term impact of some of these pesticides are still unknown.

23. Don't Buy DVDs

Some people like to buy copies of their favourite movies on DVD, but when you think about these purchases realistically, how many times do you really think you'll watch those DVDs? In most cases, DVDs are watched two to three times, at maximum. If you buy several DVDs, you'll generate waste when you eventually get rid of them. Instead, try just renting your favourite movie, and if you really want to watch it twice, rent it a second time. With the comparative cost of new and rental DVDs, you'll need to rent a movie three times before covering the price of a new one.

24. City Parks

Take advantage of your city parks. Going to the park is a great way to get away from cars and buildings and to feel the grass. Municipal parks are maintained by your tax dollars, so get involved and make your voice heard if you'd like to have playgrounds, soccer fields, or other amenities installed. Let the authorities know, and get your neighbours involved too, if need be.

25. Shorter Showers

Get in the habit of reducing your shower time. Every minute you cut can save up to five gallons of water. Over one year, this will add up to 2000 gallons of water saved per person. You can see that if you're a part of a family of five and you all make this change, this will result in a huge difference. If you have children, think about this: Do they really need to take a bath every day? You don't want to neglect hygiene, but too frequent washing causes the skin to lose its natural oils which protect against bacteria.

Web-based Resources

General Sites

www.Ecogeek.org • Small site frequently updated with articles on green topics and tips on how people can make a difference.

www.globalfootprint.org • How to calculate your organization's ecological footprint.

www.greenblue.org • This web site focuses on sustainable product design and organizational practices.

www.treehugger.com • TreeHugger is the leading media outlet dedicated to driving sustainability mainstream.

www.eatwellguide.org • The Eat Well Guide® is a free online directory for anyone in search of fresh, locally grown, and sustainably-produced food in the United States and Canada.

www.slowfood.com • The official slow food website.

www.davidsuzuki.org • The David Suzuki Foundation has worked to find ways for society to live in balance with the natural world. The Foundation uses science and education to promote solutions that conserve nature and help achieve sustainability within a generation.

Certifications and labeling

www.terrachoice-certified.com • **EcoLogo**[TM] is North America's largest, most respected environmental standard and certification mark.

Habits, good practices

www.thegreenguide.com • Very complete site with topics such as personal care, travel, kids, home and garden.

Construction

www.buildinggreen.com • If you want to learn about green construction, this site is for you.

Additional Resources

Environment

Blackburn, William R. *The sustainability Handbook: The complete management Guide to achieving Social, Economic, and environmental responsibility*. London: Earthscan, 2007.

Elkington, John. Cannibals with Forks: *The Triple Bottom Line of 21st Century Business.* Oxford: Capstone Publishing, 1997.

Esty, Daniel C, & Winston, Andrew S. *Green to gold*. N-J: John Wiley & Sons, 2006.

Friedman, Frank B. *Practical Guide to Environmental Management.* 10th ed. Washington, D.C.:Environmental Law Institute, 2006.

Holliday, Charles O., jr., Stephan Schmidheiny, and Philip Watts. *Walking the Talk: The Business Case for Sustainable Development*. Sheffield, UK: Greenleaf Publishing, 2002.

Lockwood, Charles. *Harvard Business Review on Green Business Strategy*, Harvard business school Publishing Corporation, Boston, 2007.

Lourie, B, Smith, R. *Slow Death by Rubber Duck*, Alfred A. Knoff, Canada, 2009.

Malower, J. *Strategies for the Green Economy*, Mc Graw Hill, 2009.

McDilda, D.G. *365 ways to live green*, Avon, Adams media, 2008.

Mintzer, Rich. *Green Business*, Entrepreneur media Inc, 2009.

Prakash, Aseem. Greening the firm, Cambridge, UK: Cambridge University Press, 2000.

Rogers, E, Kostigen,T. *The Green Book*, Crown Publishing group, New York, 2007.

Savitz, Andrew. *The Triple Bottom Line: How Today's Best-run Companies are Achieving Economic, Social and Environmental Success-and How You Can Too*. Hoboken, N-J: John Wiley & Sons, 2006.

Suzuki, D, Boyd, D.R. *David Suzuki's Green Guide*, Greystone books, 2008

Vasil, A. *Ecoholic: Your Guide to the Most Environmentally-friendly Information, Products and Services in Canada,* Random House Canada, 2007.

Wallace,T ,Bonin, J, McKay, K. *True Green @ Work*, National Geographic Society, 2008.

Motivation, inspiration

If you agree with the ideas in this book but always have good reasons for not starting to doing things differently, read the books below. They'll provide you with the motivation you need!

Babauta, L. *The Power of Less*, Hyperion, New York,2009.

Ferris, T. *The 4-hour Work Week*, Crown Publishers, New York, 2007

www.ingramcontent.com/pod-product-compliance
Lightning Source LLC
Chambersburg PA
CBHW022121280326
41933CB00007B/496